MILITARY SCIENCE

DRONES

BY MATT CHANDLER

TORQUE

BELLWETHER MEDIA • MINNEAPOLIS, MN

Torque brims with excitement perfect for thrill-seekers of all kinds. Discover daring survival skills, explore uncharted worlds, and marvel at mighty engines and extreme sports. In *Torque* books, anything can happen. Are you ready?

This edition first published in 2022 by Bellwether Media, Inc.

No part of this publication may be reproduced in whole or in part without written permission of the publisher. For information regarding permission, write to Bellwether Media, Inc., Attention: Permissions Department, 6012 Blue Circle Drive, Minnetonka, MN 55343.

Library of Congress Cataloging-in-Publication Data

LC record for Drones available at: https://lccn.loc.gov/2021051727

Text copyright © 2022 by Bellwether Media, Inc. TORQUE and associated logos are trademarks and/or registered trademarks of Bellwether Media, Inc.

Editor: Betsy Rathburn Designer: Jeffrey Kollock

Printed in the United States of America, North Mankato, MN.

TABLE OF CONTENTS

READY FOR TAKEOFF!	4
WHAT ARE DRONES?	6
THE SCIENCE BEHIND DRONES	12
THE FUTURE OF DRONES	18
GLOSSARY	22
TO LEARN MORE	23
INDEX	24

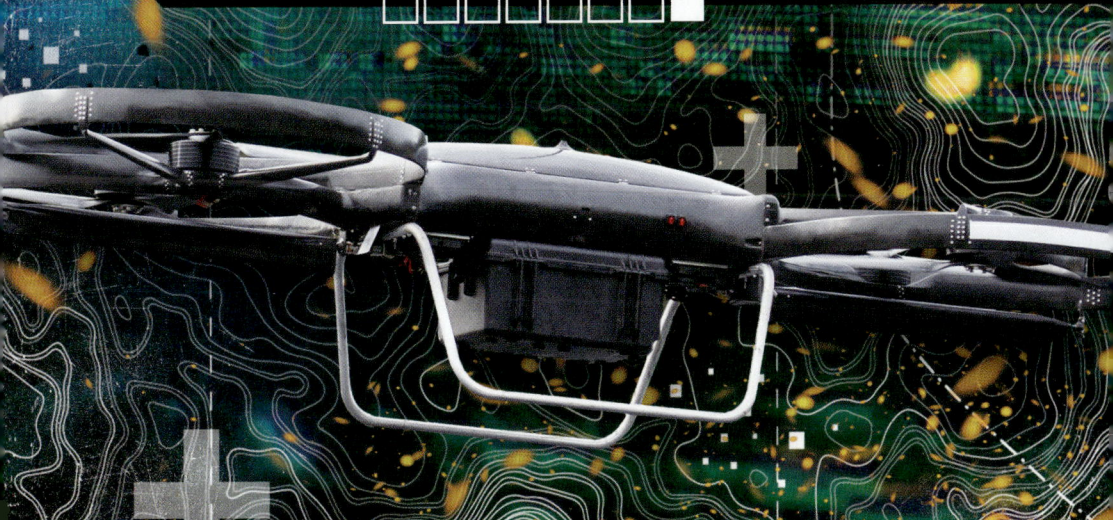

READY FOR TAKEOFF!

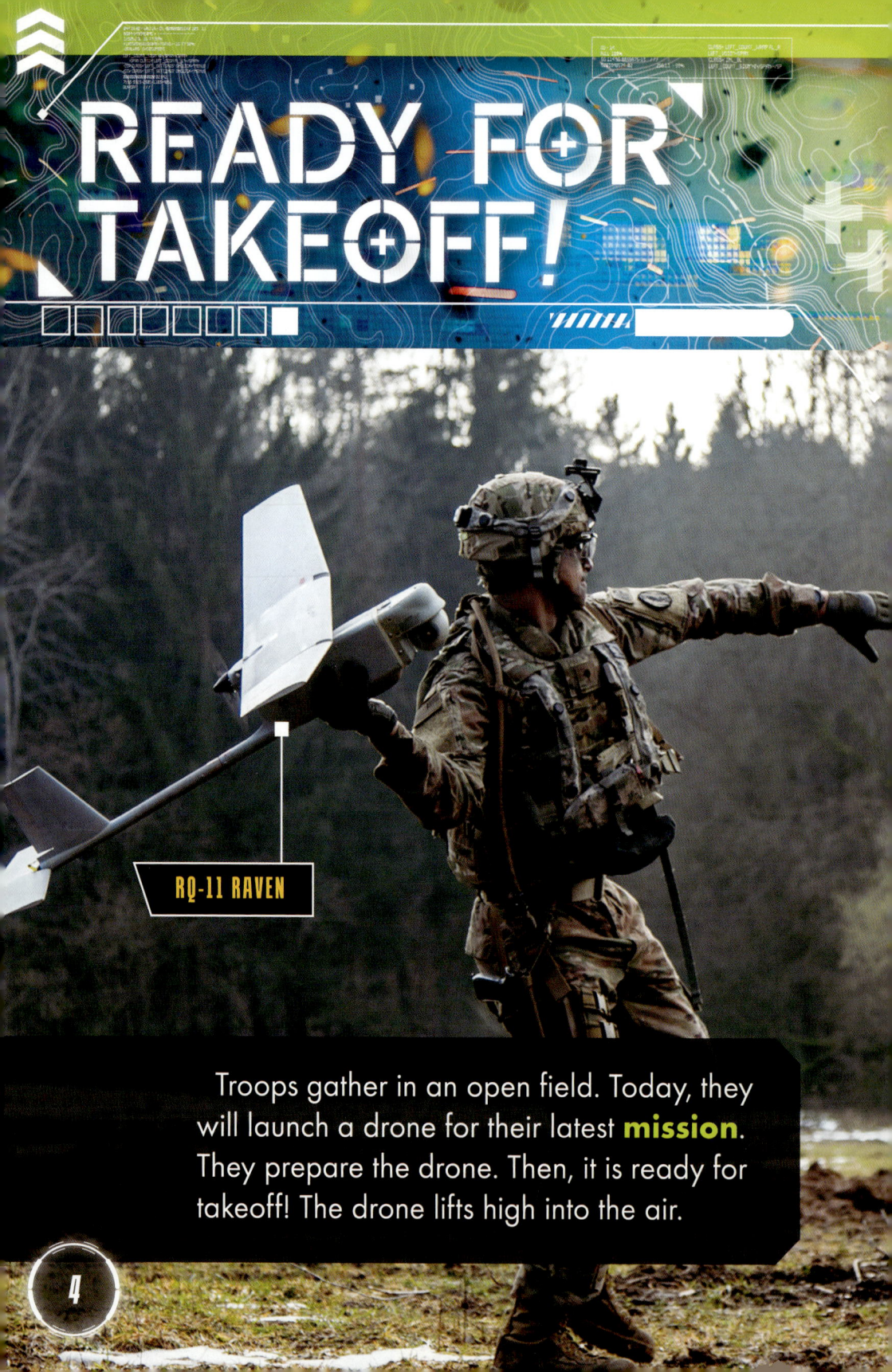

RQ-11 RAVEN

Troops gather in an open field. Today, they will launch a drone for their latest **mission**. They prepare the drone. Then, it is ready for takeoff! The drone lifts high into the air.

The mission will take the drone into enemy **territory**. It will take pictures and videos. But the troops will stay safe!

WHAT ARE DRONES?

Drones are vehicles that fly without a person inside. They come in many sizes. Small drones can fit in the palm of a hand. The largest drones look like airplanes!

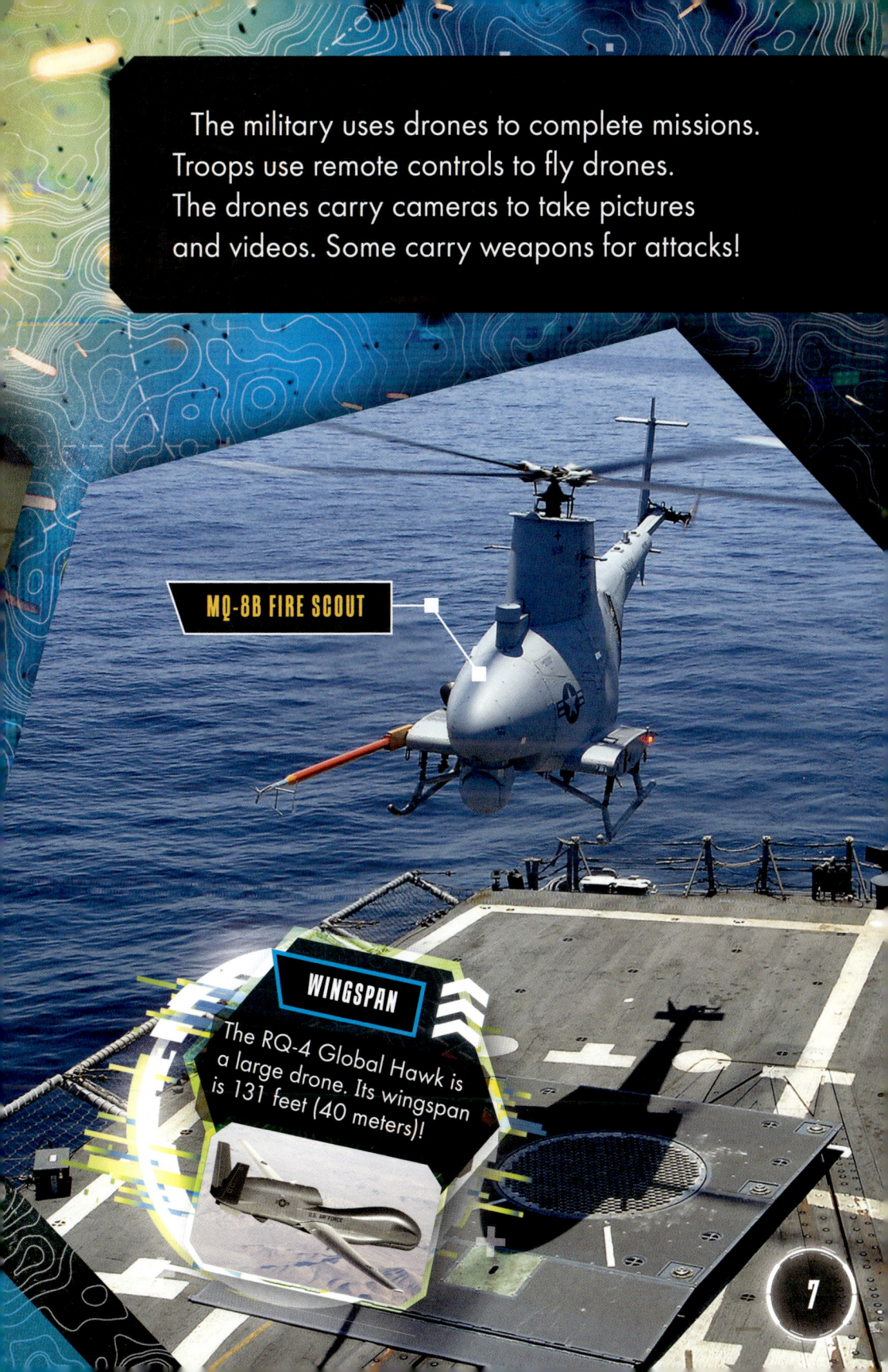

The military uses drones to complete missions. Troops use remote controls to fly drones. The drones carry cameras to take pictures and videos. Some carry weapons for attacks!

MQ-8B FIRE SCOUT

WINGSPAN

The RQ-4 Global Hawk is a large drone. Its wingspan is 131 feet (40 meters)!

The earliest drone was built in 1917. It carried almost 200 pounds (91 kilograms) of explosives. It was built to explode when it reached its target!

The first modern drone was built in 1935. It was made for **target practice**. By the 1970s, drones had cameras to send live video back to troops.

TIMELINE

1917
KETTERING BUG

1935
DE HAVILLAND QUEEN BEE

1979
IAI SCOUT

Today, drones have many purposes. Smaller drones may be used for **scouting**. They send troops videos of what lies ahead. This keeps troops safer as they travel.

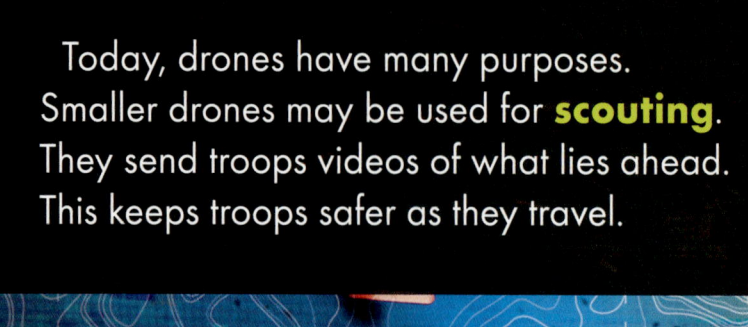

HOW FAR?

The most advanced drones can fly for days at a time. They can travel thousands of miles from their pilots!

RQ-20 PUMA

DRONE PROFILE

RQ-16A T-HAWK

RELEASED: 2007

FEATURES: SMALL DRONE THAT HAS THREE CAMERAS AND CAN TRAVEL AROUND 6 MILES (10 KILOMETERS) AWAY FROM ITS PILOT

Larger drones can fly greater distances and stay in the air longer. They are often used for spying. They can also carry bombs. Pilots may order them to fire on enemies.

THE SCIENCE BEHIND DRONES

PROPELLER

MQ-9 REAPER

Many drones look like airplanes or helicopters. They use engines and **propellers** to fly. But drones do not have pilots on board to **navigate**.

HOW DRONES WORK

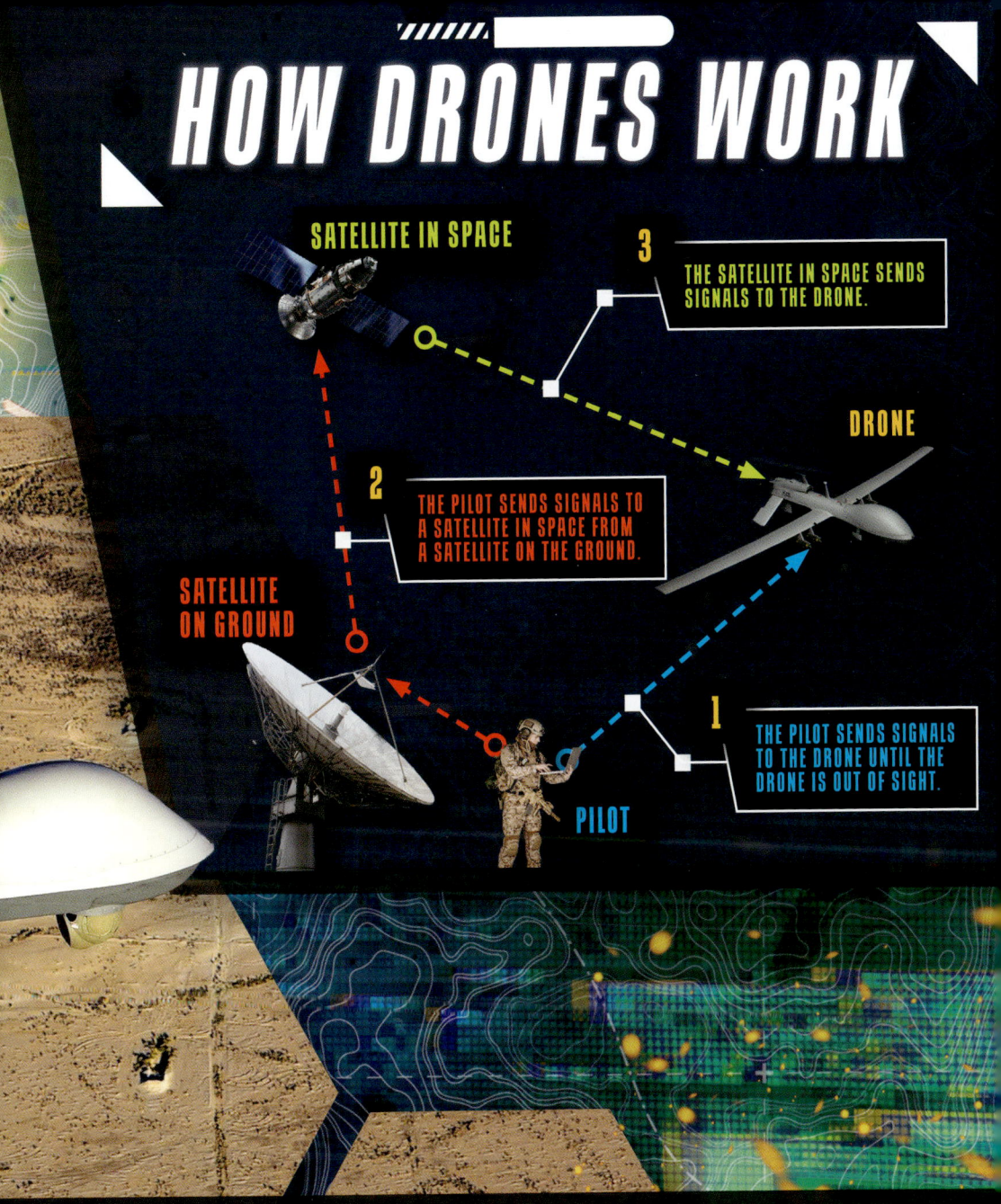

Instead, pilots use technology to control drones from far away. **GPS** devices on drones help pilots track them. Pilots can see where drones go and how far away they are from their targets.

Pilots control drones with remote controls. When drones are out of sight, pilots use **satellites**. They send **signals** to satellites. The satellites send the signals to drones as they fly. They tell the drones to turn, take pictures, and record videos.

SATELLITE

Drones send pictures and videos back to the satellites. Satellites send them to the pilots. Pilots use them to plan their next move!

MISSILE

Drones do not only take pictures. They also carry weapons! The largest drones hold **missiles** on their wings. Pilots tell them when and where to strike.

When they reach their target, the drones drop the missiles. Lasers guide the missiles to their targets.

DRONE STRIKE

THE FUTURE OF DRONES

Future drones will be faster. They will fly higher and stay in the air longer. Longer missions will let militaries gather more information.

FUTURE DRONE PROFILE

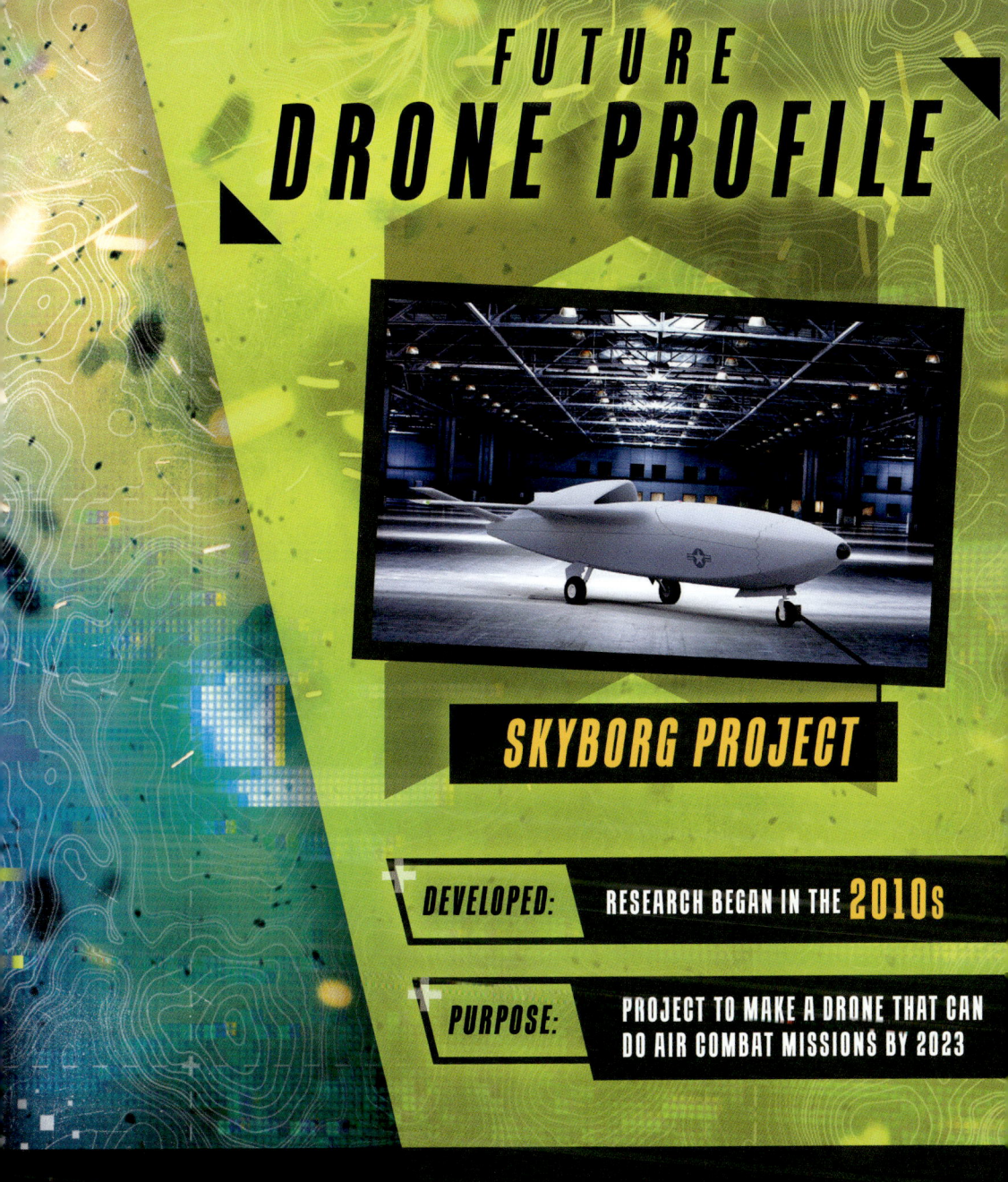

SKYBORG PROJECT

DEVELOPED: RESEARCH BEGAN IN THE **2010s**

PURPOSE: PROJECT TO MAKE A DRONE THAT CAN DO AIR COMBAT MISSIONS BY 2023

Drones may also use **AI**. This would let drones make decisions on their own. They would know where to fly. They would know when to take pictures or drop bombs!

Drone **swarms** may become common, too. Many drones are launched at once. They fly together. They send information to one another. Some will carry weapons, and others will take pictures. They will let the military do many things at once!

NON-MILITARY USES

DRONE RACING

DELIVERY

SEARCH AND RESCUE

DRONE SWARM

Future technology will make drones even more advanced. These powerful military tools are here to stay!

GLOSSARY

AI—artificial intelligence; AI is the ability of machines to copy intelligent human actions.

GPS—global positioning system; GPS is a system that uses satellites to determine locations on Earth.

missiles—powerful weapons that carry explosives

mission—a job to complete a certain task

navigate—to find the way

propellers—spinning parts that make some machines move

satellites—human-made objects that circle Earth; satellites are used to communicate and study Earth.

scouting—gathering information about enemies in the area

signals—information sent and received

swarms—groups of drones that travel together and can communicate with one another

target practice—the act of shooting at targets to improve a person's shooting abilities

territory—an area controlled by a specific person or group

TO LEARN MORE

AT THE LIBRARY

Boutland, Craig. *New Generation Vehicles: Drones, Mine Clearance, and Bomb Disposal.* North Mankato, Minn.: Capstone Press, 2019.

Noll, Elizabeth. *Surveillance.* Minneapolis, Minn.: Bellwether Media, 2022.

Rathbun, Betsy. *Drones.* Minneapolis, Minn.: Bellwether Media, 2021.

ON THE WEB

FACTSURFER

Factsurfer.com gives you a safe, fun way to find more information.

1. Go to www.factsurfer.com

2. Enter "drones" into the search box and click 🔍.

3. Select your book cover to see a list of related content.

INDEX

AI, 19
bombs, 11, 19
cameras, 7, 8
engines, 12
explosives, 8
future, 18, 19, 20, 21
GPS, 13
how drones work, 13
lasers, 17
missiles, 16, 17
mission, 4, 5, 7, 18
navigate, 12
non-military uses, 20
pictures, 5, 7, 14, 15, 16, 19, 20
pilots, 10, 11, 12, 13, 14, 15, 16
propellers, 12
remote controls, 7, 14
RQ-16A T-Hawk, 11
RQ-4 Global Hawk, 7
satellites, 14

scouting, 10
signals, 14
sizes, 6, 7, 10, 11, 16
Skyborg project, 19
spying, 11
swarms, 20, 21
target practice, 8
timeline, 8–9
troops, 4, 5, 7, 8, 10
videos, 5, 7, 8, 10, 14, 15
weapons, 7, 16, 20

The images in this book are reproduced through the courtesy of: Corporal Steve Follows RAF/ Wikimedia Commons, front cover; Pvt. Gabriel Silva/ DVIDS, pp. 2-3; Spc. Dustin Biven/ DVIDS, pp. 4-5, 5; Staff Sgt. Alan Brutus/ DVIDS, pp. 6-7 (left); Alan Gragg/ DVIDS, pp. 6-7 (right); Bobbi Zapka/ Wikimedia Commons, p. 7 (fun fact); Greg Hume/ Wikimedia Commons, p. 8 (1917); Adrian Pingston/ Wikimedia Commons, p. 8 (1935); aick/ Wikimedia Commons, p. 8 (1979); Tech. Sgt. Efren Lopez/ DVIDS, p. 9 (1995); Tech. Sgt. Emerson Nuñez/ DVIDS, p. 9 (2007); DVIDS, pp. 9 (2018), 11, 19; Michael Farmer/ Alamy, pp. 8-9; NASA/ Alamy, p. 10 (fun fact); Sgt. Manuel Serrano/ DVIDS, pp. 10-11; Airman 1st Class William Rosado/ DVIDS, pp. 12-13; Andrey Armyagov, p. 14; Staff Sgt. Travis Fontane/ DVIDS, p. 15; 1st Lt. Aaron DeCapua, p. 16 (inset); Staff Sgt. Brian Ferguson/ DVIDS, p. 16; SugaBom86, p. 17; Cpl. Rachel Young-Porter/ DVIDS, pp. 18-19; Drone photography, p. 20 (racing); The Toidi, p. 20 (delivery); Sobrevolando Patagonia, p. 20 (rescue); Frame Stock Footage, pp. 20-21; Andy Dean Photography, p. 21 (inset); aapsky, p. 23.